AVA THE S.T.E.M. PRINCESS®

This book is dedicated to my Mom and Dad, Tita and Terrence Simmons, my sister, Chynah Jeter, my grandma, Mary L. Foy, my cousins, Yevette Holmes and LaKecia Holmes, and my sweet puppy, Shelby, who encourage me daily to be the Genius that I am. Extra special thank you to my grandma for initiating the S.T.E.M. journey for our family.

Thank you to the S.T.E.M. Leaders, Team Genius Squad Board members, and key contributors who believed in me, encouraged me, and supported the mission and vision of Team Genius Squad.

Thank you to all of the Geniuses around the world that support Team Genius Squad, I truly appreciate you for all of your support!

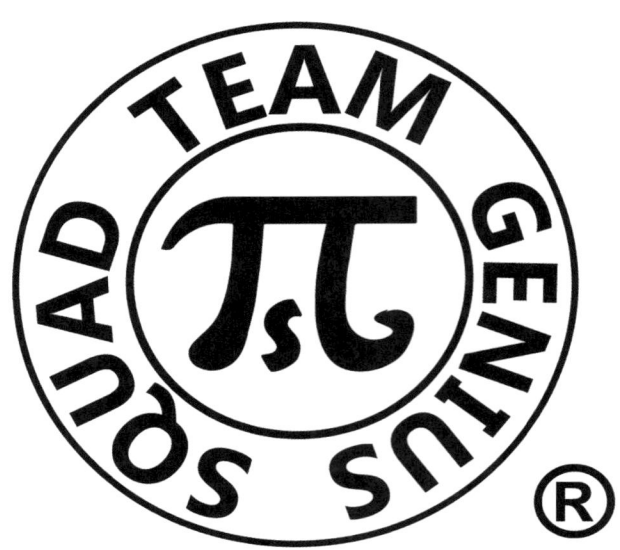

Team Genius Squad Publishing

This coloring and activity book was authored, including original art and page scenes, by Ava N. Simmons. It is also aligned with the International Society for Technology in Education (ISTE) standards, state-level Science Standards, and the 5E Model (Engage, Explore, Explain, Elaborate, and Evaluate).

ISBN: 979-8-9860155-0-7
Book Type: Children's, Family


All inquiries regarding this book can be sent to the author at info@TeamGeniusSquad.com. For more information about the author or Team Genius Squad, please visit www.TeamGeniusSquad.com

I AM A GENIUS!

1

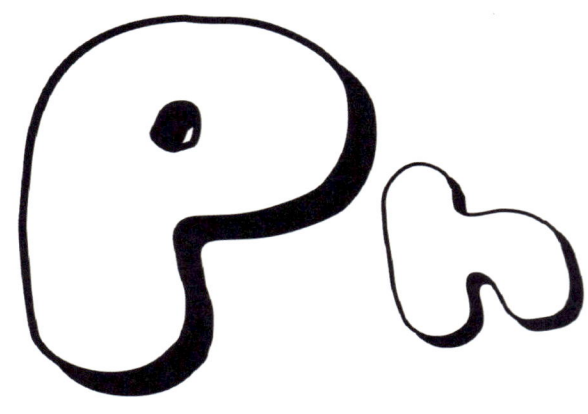

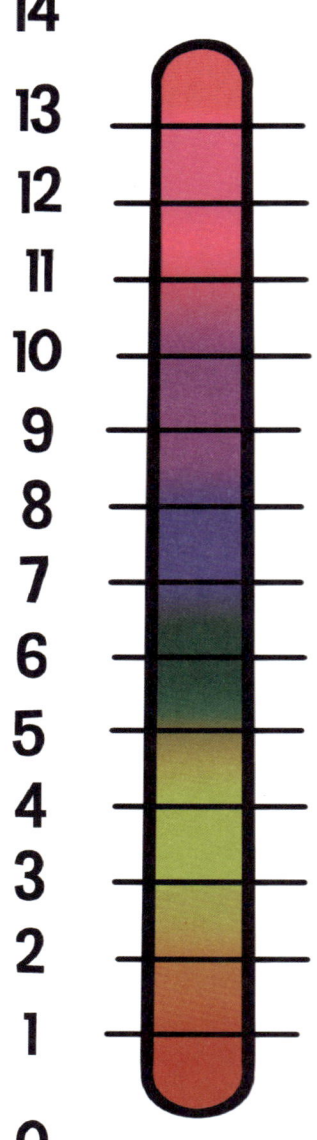

| | |
|---|---|
| 14 | |
| 13 | BLEACH |
| 12 | SOAPY WATER |
| 11 | AMMONIA SOLUTION |
| 10 | MILK OF MAGNESIA |
| 9 | BAKING SODA |
| 8 | SEA WATER |
| 7 | DISTILLED WATER |
| 6 | URINE |
| 5 | BLACK COFFEE |
| 4 | TOMATO JUICE |
| 3 | ORANGE JUICE |
| 2 | LEMON JUICE |
| 1 | GASTRIC ACID |
| 0 | |

**MORE BASIC**

**MORE ACIDIC**

pH means potential hydrogen and ranges from 0-14. pH is used to measure the acidity and alkalinity of substances. When the pH is 0-6 it's an Acid, when the pH is 7 it's neutral, and when the pH is 8-14 it's a Base. Acids and Bases can be foods, chemicals or body fluids.

# ACIDS

VINEGAR

LEMON

# ACIDS HAVE A PH OF 0 - 6

Vinegar and Lemons are Acids. Vinegar has a pH of 3 and Lemons have a pH of 2. This means Lemons are more acidic than Vinegar.

# ACIDS

TOMATO

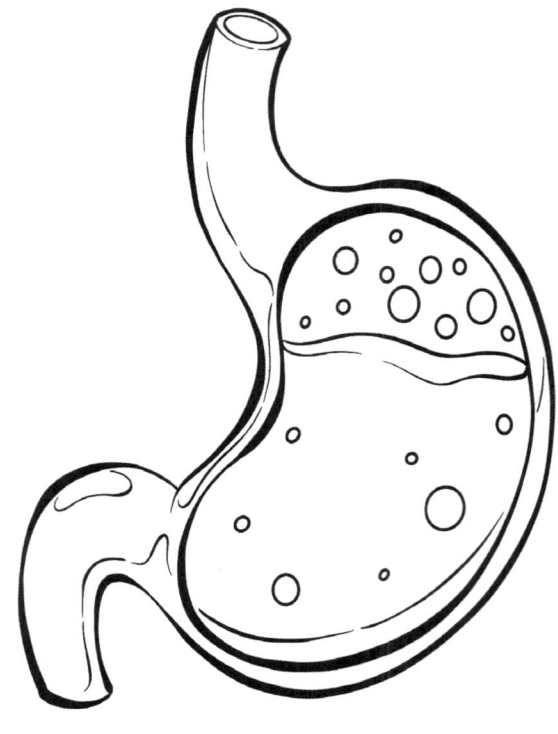

STOMACH ACID

ACIDS HAVE A PH OF 0 - 6

Your stomach makes Acid and a Tomato is an Acid. Tomatoes are less acidic than Vinegar and Lemons but Stomach Acid is more acidic than all of them.

# ACID AND THE BODY

## LEMON HAS A PH - 2 AKA ACIDIC

Acidic foods taste tart and sour. When you bite a Lemon it tastes tart and sour.

# FUN S.T.E.M. MAZE

**Drive the rocket through the maze to reach earth!**

# BASES

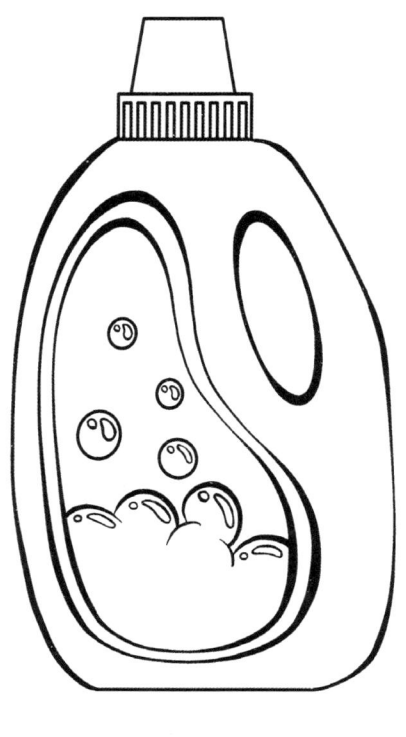

## HOUSEHOLD CLEANER

## BAKING POWDER

# BASES HAVE A PH OF 8-14

Household Cleaner and Baking Powder are Bases. Household Cleaner has a pH of 11 and Baking Powder has a pH of 8.3 to 9. Solutions and powders with a pH greater than 7 are great for removing dirt and grease.

# BASES

MILK

BAKING SODA

## BASES HAVE A PH OF 8-14

Milk and Baking Soda are Bases. Milk has a pH of 6.7 and Baking Soda has a pH of 8.3 to 9. Depending on the pH of an Acid or a Base, mixing them together can cause a calm reaction or a bubbly explosive reaction.

# BASE AND THE BODY

TUMMY ACHE / STOMACH ACID

MILK

PEPTO BISMOL

Pure Baking Soda

## BASES CAN CALM ACIDS

Too much Stomach Acid can cause a tummy ache. Mixing Bases with Acids can neutralize Acids and reduce their acidity. For example, when you drink milk or take baking soda type products it can make an upset acidic stomach

# FUN ACID-BASE
## Crossword Puzzle

```
E A Y J Z X S D S V F T C J T I E
E D U R B T S T R I X Z Y Q J L Z
Y C I Z E K B B M N Y R J F T U J
B A S E H A W E Z E S K H P P W O
A E C H H Q C U K G P Y D Q U I J
O F Z I C K N T T A I S J F P I D
T I O K J R Z M I R Y U K Z W B P
B C T O K L C H Z O C V A C I D E
A K X M D B S G Z G N B Z K C S I
K M D F E C B U B B L E S I R G Z
I Y I Q R T O X V A F S L W C Y R
N Y S Q G V S L R P L O E Y T J T
G D H H H X F T O W P W Q B Q N D
S S S Z J Y U Y X R R J I Q U B M
O S O U O E T N U V I Q P F H Q Z
D B A O N U Q N I I S N E W Z J X
A G P C I W C H D A N X G T F G J
```

| | | |
|---|---|---|
| FUN | BAKING SODA | FOOD COLORING |
| VINEGAR | DISH SOAP | NEUTRAL |
| BUBBLES | STEM | PH |
| REACTION | BASE | ACID |

# NEUTRAL

POT OF WATER

BUCKET OF WATER

## WATER HAS A NEUTRAL PH OF 7

Water has a pH of 7 making it a neutral solution. Neutral solutions mix well with acids and bases.

# NEUTRAL

GLASS OF WATER          BOTTLED WATER

# WATER HAS A NEUTRAL PH OF 7

The pH of substances can create a strong reaction when mixed together. For example if you add water, pH 7, to a strong acid, pH 1, it will react quickly but if you add a strong acid to water the reaction will be slower.

# WATER AND THE BODY

## WATER HAS A NEUTRAL PH OF 7

Water has no taste and is helpful to the body because the pH is neutral. Adding Acid or Base to water can decrease or increase the pH of water. If you add a Base to water it becomes more basic and is called a high pH water.

# FUN PH GAME

**Place The Stickers Under The Correct pH Level**
**(Hint: Use the pH Chart)**

## ACID
0-6

## NEUTRAL
7

## BASE
8-14

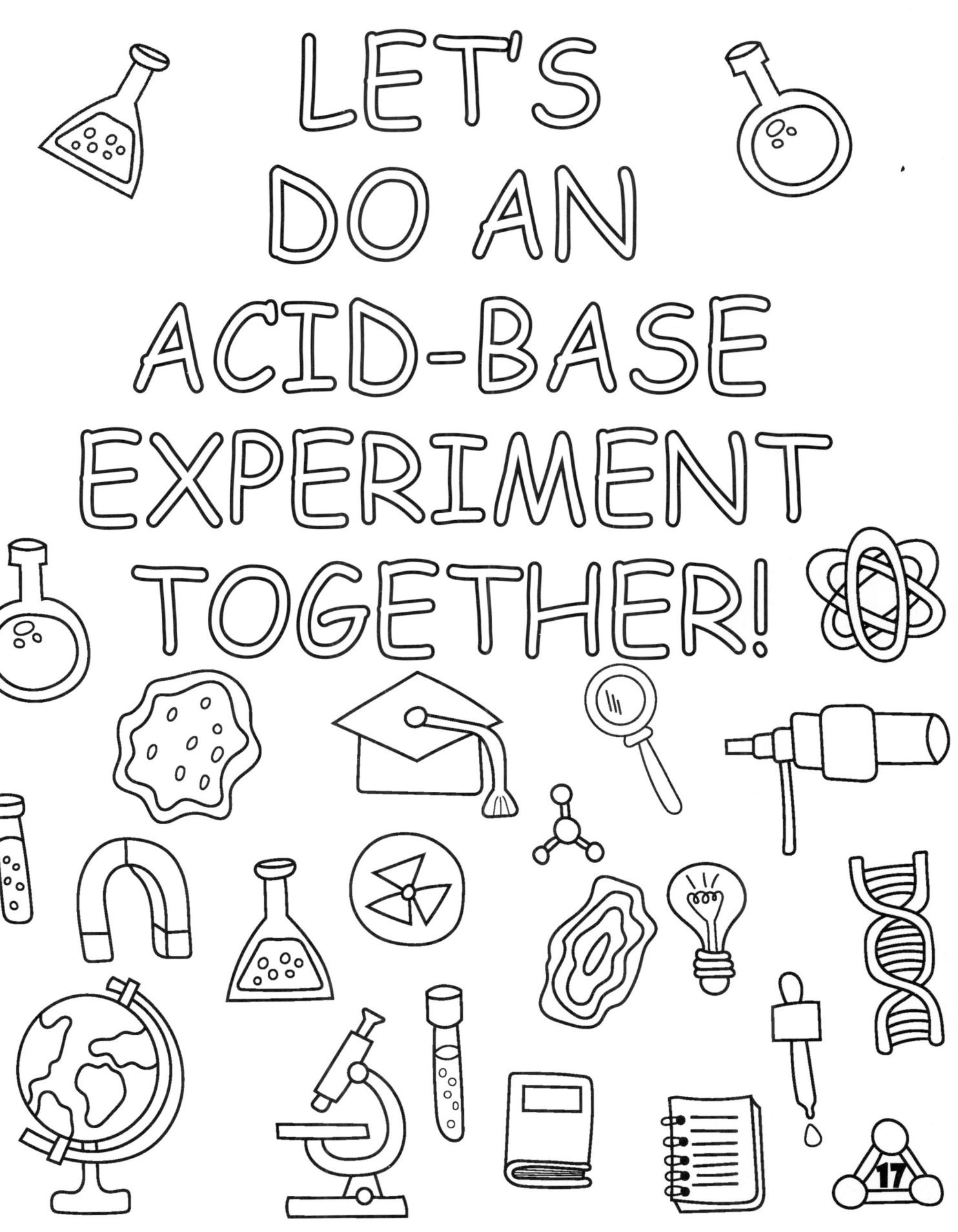

# INGREDIENTS

DISH SOAP

VINEGAR

FOOD COLORING

BAKING SODA

**INGREDIENTS:**
2 Cups Vinegar (aka Acetic Acid, pH 3)
1 Tablespoon of Dish Soap (aka Surfactant, pH 12)
1 Teaspoon of Food Coloring (pH 10)
2 Tablespoons of Baking Soda (aka Sodium Bicarbonate, pH 8.3-9)

# ADULT SUPERVISION

TABLE OF ELEMENTS

Ensure to have adult supervision when doing your experiment. An adult can be your Lab Assistant and join in on the fun!

# TIME TO EXPERIMENT!

Now you are ready to do the experiment! Add your ingredients one at a time to your mixing container. Pay attention to the reactions. Ensure to add the Baking Soda last and be ready for the fun reaction!

# ADD THE INGREDIENTS IN ORDER FOR THE BEST REACTION!

1. VINEGAR
2. DISH SOAP
3. FOOD COLORING
4. BAKING SODA

# THEN!!!!!

REACTION FUN!

22

NOW WATCH
AVA THE S.T.E.M. PRINCESS®

DO THE ACID-BASE REACTION

EXPERIMENT! YOU CAN WATCH HER

VIDEO BY SCANNING THIS QR CODE:

OR BY VISITING HER YOU TUBE LINK:

https://youtu.be/Y9MvlCqA4_s

# FUN EXPERIMENT GAME!

**Match The Ingredients With the Experiment Order.**
**Draw A Line Between The Ingredient And Correct Order Number.**

**(Hint: Check The Experiment Order You Used In The Experiment)**

1.      Baking Soda

2.      Vinegar

3.      Dish Soap

4.      Food Coloring

# EXPLORE MORE

**Let's Try Other Liquids With The Experiment To See What Happens!**

1. Use The Ingredients:
   - 2 Cups of Coffee (pH 5)
   - 1 Tablespoon of Dish Soap (pH 12)
   - 1 Teaspoon of Food Coloring (pH 10)
   - 2 Tablespoons of Baking Soda (pH 8.3 - 9)

What happened?_____

How long before something happened?_____

2. Use The Ingredients:
   - 2 Cups of Orange Juice (pH 3)
   - 1 Tablespoon of Dish Soap (pH 12)
   - 1 Teaspoon of Food Coloring (pH 10)
   - 2 Tablespoons of Baking Soda (pH 8.3 - 9)

What happened?_____

How long before something happened?_____

3. Use The Ingredients:
   - 2 Cups of Water (pH 7)
   - 1 Tablespoon of Dish Soap (pH 12)
   - 1 Teaspoon of Food Coloring (pH 10)
   - 2 Tablespoons of Baking Soda (pH 8.3 - 9)

What happened?_____

How long before something happened?_____

# OBSERVATIONS

**Let's Review and Analyze What Was Observed!**

1. Were there differences in what happened when you used different ingredients?

_____

_____

2. Describe the differences you saw.

_____

_____

3. Why do you think this happened?

_____

_____

4. Do Research To Confirm Your Answer. Ask an Adult or with an Adult look for information in books or on the internet.

_____

_____

5. What is your final conclusion on the experiment reaction when liquids with different pHs are used?

_____

_____

# MORE FUN S.T.E.M. COLORING PAGES!

**Example S.T.E.M. Careers & Areas**

# CHEMIST
## Works With Chemicals and Analyzes The Way They React

# ENGINEER

Designs and Builds Machines or Structures

# MATH TEACHER

## Teaches Math Concepts To Others

# CODER

## Writes Codes To Make Computers Work

# ASTRONAUT
## Works in Outer Space aboard a Spacecraft

# AEROSPACE - ASTRONAUTICS

**The Study of Aircraft Design & Flight Technology in Outer Space**

33

# ROBOTICS
The Study of the Design, Operation, & Use of Robots

# ASTRONOMY
## The Study of the Sun, Moon, Planets, and Stars

EARTH

SUN

VENUS

MOON

SATURN

MERCURY

NEPTUNE

MARS

JUPITER

# FAMILY FUN EXPERIMENTS!

Experiments are a great family fun activity. Ensure to share this fun experiment with your family!

# CERTIFIED GENIUS!

| 32 | 28 | 92 | 16 |
|---|---|---|---|
| **Ge** | **Ni** | **U** | **S** |
| Germanium | Nickel | Uranium | Sulfur |
| 72.631 | 58.693 | 238.029 | 32.066 |

SQUAD!!!

You have completed the experiment so you are now a
Team Genius Squad Certified Genius! Congratulations!

# NEXT EXPERIMENT
# LEMON ELECTRICITY
# STAY TUNED...

**Stay tuned for our next coloring and activity book that will cover the Lemon Electricity Experiment. It's so exciting and I know you will love it!**

# AVA N. SIMMONS
# AVA THE S.T.E.M. PRINCESS®

S.T.E.M. Ambassador, Entrepreneur, Influencer, Designer, Author, Inspirational Leader

Thank you for taking the time to do an experiment with me. You are now a member of Team Genius Squad!

Remember that you have purpose and unique talents; so, when challenges come your way focus on those great things within you. I found my inner Genius by focusing on my strengths and not letting my challenges shape my future; I want to encourage you to do the same.

I believe in you and can't wait for you to make your unique mark on the world!

Your Fellow Genius,

*Ava N Simmons*

Ava N. Simmons
CEO, Team Genius Squad

Ava The S.T.E.M. Princess® is building a Mobile S.T.E.M. Lab Truck. Currently, her Team Genius Squad provides interactive S.T.E.M. Lab activities in the community via an on-site canopy tent. Due to overwhelming demands, she will now expand the community S.T.E.M. Lab to a Lab on wheels. We thank everyone for their continued support with this monumental project!

**COMING SOON!**

Fun S.T.E.M. Experiments on wheels!

# STAY CONNECTED WITH TEAM GENIUS SQUAD

## Websites

Website: www.TeamGeniusSquad.com
Store: www.Shop.TeamGeniusSquad.com

## The Team Genius Squad APP

http://TeamGeniusSquadConnect.com/

## Social Media

♦ Instagram: @AvaTheSTEMPrincess
♦ Facebook: @AvaTheSTEMPrincess
♦ Twitter: @AvaTheSTEMPrin1
♦ LinkedIn: @AvaTheSTEMPrincess
♦ LinkedIn: @TeamGeniusSquad
♦ YouTube: https://YouTube.com/c/AvaTheSTEMPrincess

## Mailing Address

Team Genius Squad
8311 Brier Creek Parkway, Suite 105-260
Raleigh, North Carolina, 27703 USA
Tel: 802-277-0332
Email: info@TeamGeniusSquad.com

# CROSSWORD PUZZLE
## ANSWER KEY

```
E A Y J Z X S D S V F T C J T I E
E D U R B T S T R I X Z Y Q J L Z
Y C I Z E K B B M N Y R J F T U J
B A S E H A W E Z E S K H P P W O
A E C H H Q C U K G P Y D Q U I J
O F Z I C K N T T A I S J F P I D
T I O K J R Z M I R Y U K Z W B P
B C T O K L C H Z O C V A C I D E
A K X M D B S G Z G N B Z K C S I
K M D F E C B U B B L E S I R G Z
I Y I Q R T O X V A F S L W C Y R
N Y S Q G V S L R P L O E Y T J T
G D H H H X F T O W P W Q B Q N D
S S S Z J Y U Y X R R J I Q U B M
O A O U O E T N U V I Q P F H Q Z
D B A N U Q N I I S N E W Z J X
A G P C I W C H D A N X G T F G J
```

# FUN EXPERIMENT GAME
## ANSWER KEY

**Match The Ingredients With the Experiment Order.**
**Draw A Line Between The Ingredient And Correct Order Number.**

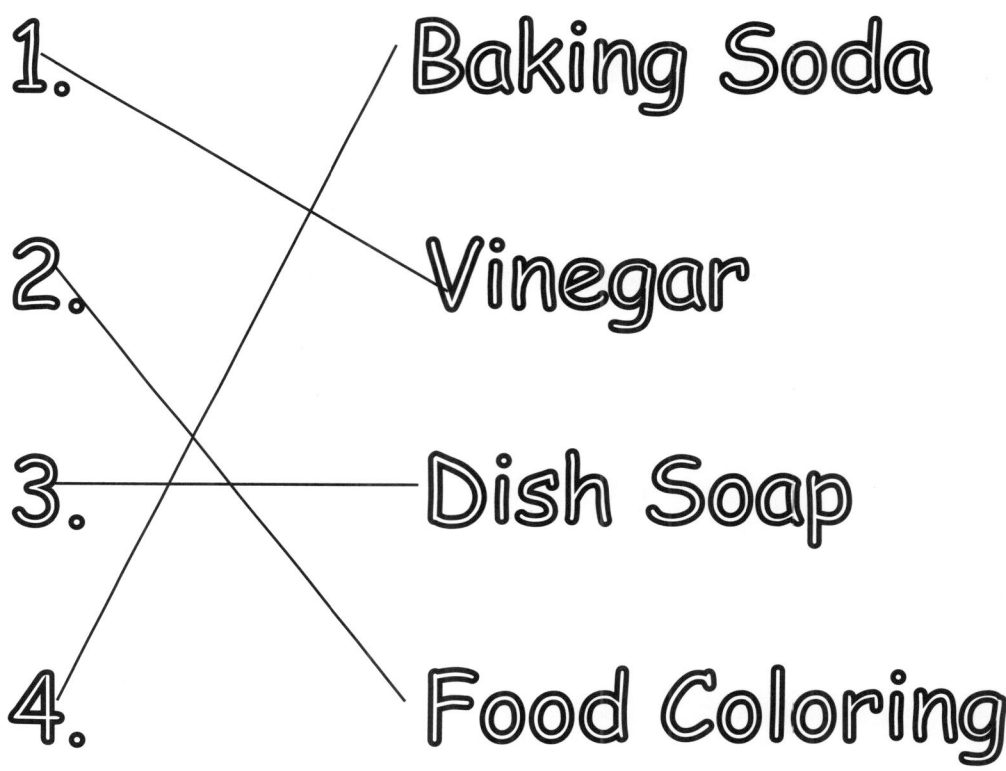

1.

2.

3.

4.

Baking Soda

Vinegar

Dish Soap

Food Coloring

# FUN PH GAME
# ANSWER KEY

**Place The Sticker Item Under the Correct pH Level**
**(Hint: Use the pH Chart)**

## ACID
### 0-6

## NEUTRAL
### 7

## BASE
### 8-14